Busy Bugs

FLIES

By Bray Jacobson

Please visit our website, www.garethstevens.com. For a free color catalog of all our high-quality books, call toll free 1-800-542-2595 or fax 1-877-542-2596.

Library of Congress Cataloging-in-Publication Data
Names: Jacobson, Bray, author.
Title: Flies / Bray Jacobson.
Description: New York : Gareth Stevens Publishing, [2022] | Series: Busy bugs | Includes index.
Identifiers: LCCN 2020008133 | ISBN 9781538263358 (library binding) | ISBN 9781538263334 (paperback) | ISBN 9781538263341 (6 Pack) | ISBN 9781538263365 (ebook)
Subjects: LCSH: Flies–Juvenile literature.
Classification: LCC QL533.2 .J33 2022 | DDC 595.7/35–dc23
LC record available at https://lccn.loc.gov/2020008133

First Edition

Published in 2022 by
Gareth Stevens Publishing
111 East 14th Street, Suite 349
New York, NY 10003

Editor: Kristen Nelson
Designer: Katelyn E. Reynolds

Photo credits: Cover, p. 1 MR.AUKID PHUMSIRICHAT/Shutterstock.com; p. 5 Arterra/Universal Images Group via Getty Images; pp. 7, 24 (eggs) lensblur/ iStock / Getty Images Plus; pp. 9, 24 (larva) Gustavo Mazzarollo/ Moment / Getty Images Plus; p. 11 Goran Safarek / EyeEm/Getty Images; p. 13 Anthony Bannister/ Gallo Images / Getty Images Plus; pp. 15, 24 (pupa) George D. Lepp/ Corbis Documentary/Getty Images; p. 17 BirdShutterB/ iStock / Getty Images Plus; p. 19 Luis Castaneda Inc./ The Image Bank / Getty Images Plus; p. 21 Roel_Meijer/ iStock / Getty Images Plus; p. 23 Photographed and edited by Janos Csongor Kerekes/ Moment/Getty Images.

Printed in the United States of America

CPSIA compliance information: Batch #CSGS22: For further information contact Gareth Stevens, New York, New York at 1-800-542-2595.

Contents

Get to Know Flies 4

Look at Larvae 8

Grown Up 16

Fly Bite 22

Words to Know 24

Index. 24

Flies live all over
the world.
There are many kinds.

Flies lay eggs.
There can be hundreds!

Larvae come out of
the eggs.
They look like worms.

They like wet places.

They eat a lot.
They grow and shed skin.

They rest as pupae.
Their bodies change.

Adult flies all have wings. They have three pairs of legs.

They are out during the day.

Most get food from flowers.

Some flies bite!

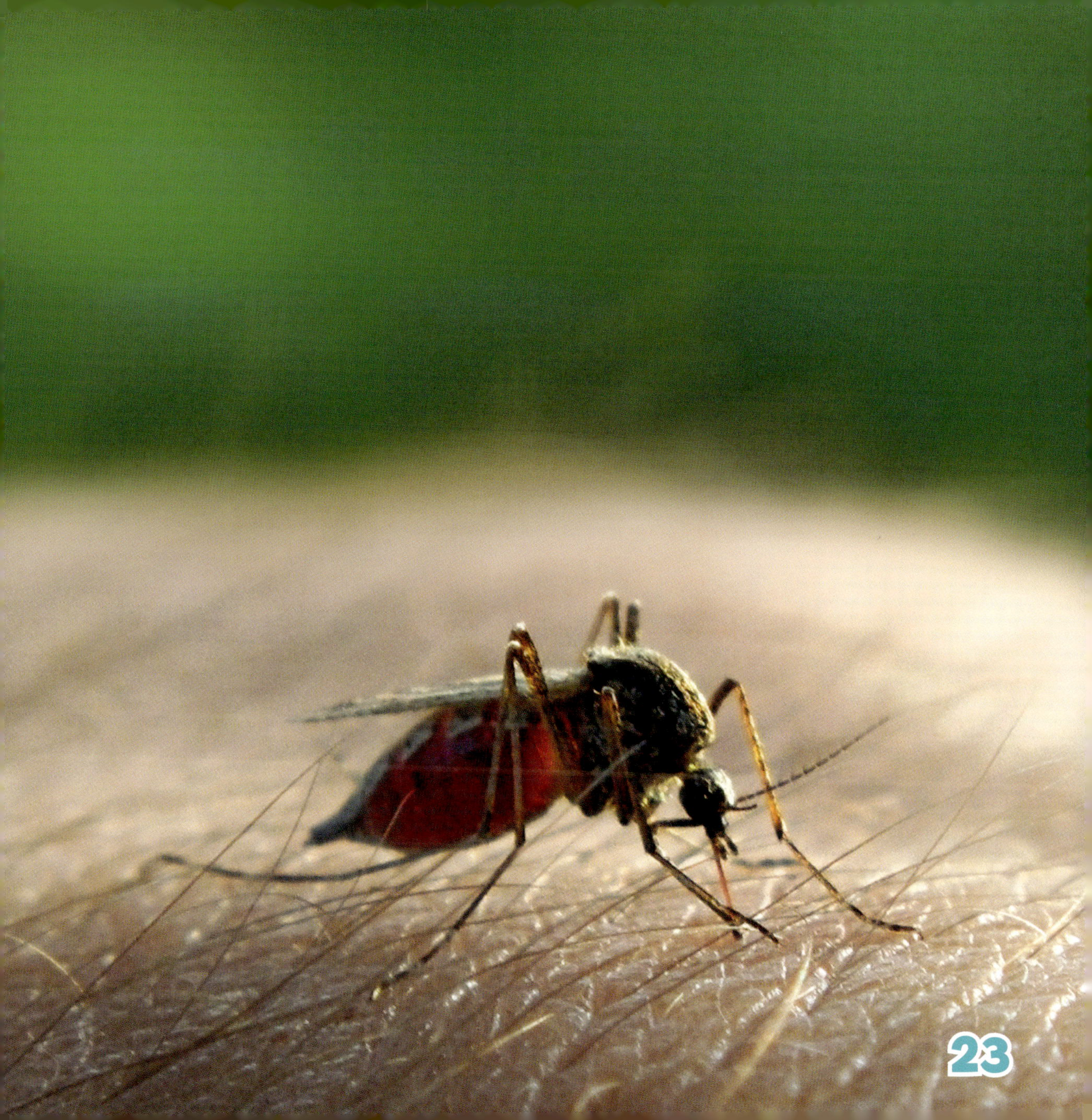

Words to Know

eggs

larva

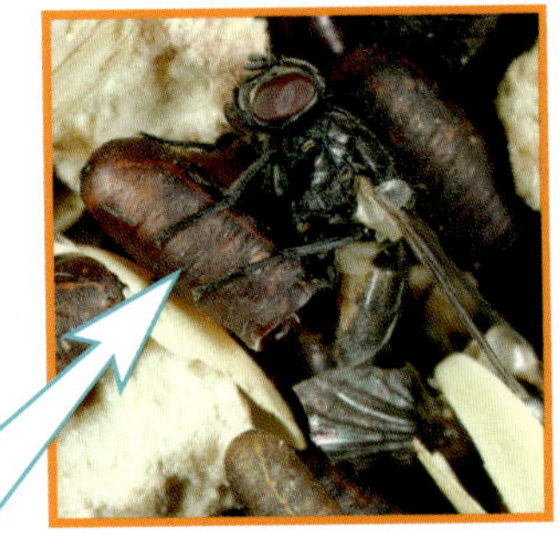
pupa

Index

bite 20

food 20

legs 16

wings 16